Joachim Bonkoungou

Farming practices and techniques

AF376418

Joachim Bonkoungou

Farming practices and techniques

Adaptation to staggered rainy season onset, Commune Rurale de Gbomblora, South-West region, Burkina Faso

ScienciaScripts

Imprint

Any brand names and product names mentioned in this book are subject to trademark, brand or patent protection and are trademarks or registered trademarks of their respective holders. The use of brand names, product names, common names, trade names, product descriptions etc. even without a particular marking in this work is in no way to be construed to mean that such names may be regarded as unrestricted in respect of trademark and brand protection legislation and could thus be used by anyone.

Cover image: www.ingimage.com

This book is a translation from the original published under ISBN 978-620-6-72604-3.

Publisher:
Sciencia Scripts
is a trademark of
Dodo Books Indian Ocean Ltd. and OmniScriptum S.R.L publishing group

120 High Road, East Finchley, London, N2 9ED, United Kingdom
Str. Armeneasca 28/1, office 1, Chisinau MD-2012, Republic of Moldova, Europe
Printed at: see last page
ISBN: 978-620-8-29354-3

Copyright © Joachim Bonkoungou
Copyright © 2024 Dodo Books Indian Ocean Ltd. and OmniScriptum S.R.L publishing group

Agricultural practices and techniques: adapting to the staggered onset of the rainy season, Gbomblora rural community, South-West region, Burkina Faso

BONKOUNGOU Joachim,
Senior Research Fellow in Geography,
Climate and the Environment
CNRST/INERA/CREAF Kamboinsé
Burkina Faso

Table of contents

Introduction

Africa is the continent most vulnerable to climate change (Baiou, 2008). And yet, it is only a small contributor to greenhouse gas emissions, a major cause of global warming. The Intergovernmental Panel on Climate Change (IPCC) predicts that the African economy will be negatively impacted, even if development opportunities can be seized (IPCC, 2014). This is particularly true of the green economy, an essential pathway for containing global warming to acceptable proportions (United Nations - Reducing Emissions from Deforestation and forest Degradation (UN-REDD), 2015). The Paris Agreement set a threshold of 1.5°C of warming that must not be exceeded (United Nations Framework Convention on Climate Change, 2015).

This contribution is set to increase as a result of practices and techniques that encourage the emission of methane, a greenhouse gas with a warming potential at least twenty-seven (27) times greater than that of carbon. (Naqvi & Sejian, 2011). Methane is emitted by rice cultivation and bush fires, among other things. African cities, which are major consumers of rice, have forced the development of rice growing. This is the case, in West Africa, with the Regional Support Programme for the Sahel Irrigation Initiative, covering six (6) Sahel countries (Tientiga et al., 2024).

In order to adapt sustainably while reducing greenhouse gas emissions, the IPCC recommended and put in place mechanisms to help low-income countries plan low-carbon development. Development plans that have already been drawn up should incorporate actions to adapt to and mitigate climate change in order to be more resilient. National adaptation programmes of action were to accompany these plans. These will be replaced by national adaptation plans and nationally determined contributions (UNFCCC, 2021). Mitigation efforts have been increased to ensure that the threshold set by the Paris Agreement is not exceeded.

Despite these efforts, hopes of limiting global warming to 1.5°C are likely to be exceeded, given the current level of 1.3°C. This means that Africa's economies will have to cope with increasingly violent and frequent climatic extremes. Yet its economy is based on agriculture, a highly vulnerable sector (Ouédraogo et al., 2022).

It was with this in mind that African countries undertook to finance agriculture to the tune of 10% of their gross domestic product (GDP) (EU-AU, 2008). Unfortunately, only a few countries, including Burkina Faso, have honoured this commitment. As a result, agriculture has not been able to act as a lever for national economies.

Despite this investment, agriculture in Burkina Faso has not developed in such a way that it is not dependent on climatic factors. National agricultural production certainly depends on increasing the area sown, which cannot be expanded at will (AGRECO, 2006). Climatic extremes, particularly the many pockets of drought and flooding, reduce and sometimes cancel out yields (Sultan et al., 2012). Food needs are covered by massive imports of rice in particular.

There are two types of agriculture in Sahelian countries. Off-season agriculture, a practice of adapting to climate change, is practised during the dry season. Rain-fed agriculture is also practised during the rainy season. It concerns all farmers and all fields (Slimane, 2008). Off-season agriculture is practised around water points.

The length of the rainy season decreases with latitude. The more southerly regions are the first to receive the monsoon winds and the last to see them descend further south. As a result, their rainy seasons last at least six (6) months. On the other hand, the northern zones are the last to receive the rains and the first to lose them. This is why the Dori region has a rainy season lasting no more than three months. Local crop cycles were therefore adapted. Early in the north and late in the south, where the season allows two harvests for early varieties. The north cannot afford to sow long-cycle crops, as the length of time they take to ripen does not allow them to mature properly.

1. Issues and objectives

The growing season can be divided into three main phases: installation, peak season and the end of the rainy season. The end of the rainy season is generally short. It occurs in September or later, depending on the climatic zone. In the Sahel, September marks the end of the season, with the last useful rains coming in mid-September at the earliest and late September at the latest. In the Sudano-Sahelian climate, the end of the season occurs from the last dekad of September to the first dekad of October. In the Sudanese climate, the end of the season occurs in October and sometimes November.

The peak rainy season is July and August, when most of the rain falls. August is by far the wettest month. Cereal crops generally ripen in late August/early September, when the high rainfall is favourable. After the last rains, out-of-season rains sometimes occur between November and April. These can cause major damage and considerable crop losses.

The rainy season generally lasts from April to June, even July and August in years with poor rainfall. This covers the sowing period. According to some authors, farmers make choices. Cycle crops are sown first and early crops last.

One of the major constraints of rainfed agriculture is the onset of the rainy season (Diallo, 2010). The literature on this subject tells us of a long and capricious installation that leads to numerous reseeding operations (Bonkoungou, 2007; Hauchart, 2007; Kambiré, 2023; Sultan et al., 2012).. Many growers lose their seeds and rely on the solidarity of others to obtain them. It is not uncommon to see seed being sown in July and even early August, the two wettest months for meeting crop water requirements.

1.1. Research questions and hypotheses

The main research question is what is the state of play at the start of the rainy season in the rural commune of Gbomblora, located about 20 km south of Gaoua, the capital of the Poni Province and the South-West Region of Burkina Faso. The aim is to provide answers to two secondary questions:

- What farming practices and techniques do farmers apply at the start of the crop year?

- What are the main constraints they face?

This is a highly contrasted situation, where farmers, guided by the desire to ensure sufficient food for their families, are faced with a busy sowing calendar. To do this, they primarily use their local know-how, which they combine with modern means of production. They are helped in this by national and local development policies that place greater emphasis on physical achievements that are beyond the reach of farmers. Capacity-building for rural farming households remains marginal. Yet these are the core of farming, and the focus should be on them too.

1.2. Objectives and expected results

The overall aim of this document is to analyse the start of the rainy season in the rural commune of Gbomblora, in the South-West region of Burkina Faso. It is part of a general context of controlling the start of the season, when farmers have busy calendars. The specific aims are to :

- Analyse the practices and techniques implemented during the installation of the growing season and

- Analyse the constraints faced by producers.

The expected result is the drafting of this document, which comprises three essential parts:

- Materials and methods;

- Results on practices and techniques used ;

- Constraints encountered.

2. Materials and methods

The study was carried out in the rural commune of Gbomblora, which we will present briefly below. It is located in the South-West region, which has four provinces. From north to south are the province of Ioba, whose capital is Dano, the province of Bougouriba, whose capital is Diébougou, the province of Poni, whose provincial capital is Gaoua, and the province of Noumbiel, whose capital is Batié. The region is bordered to the west by the Cascades and Hauts-Bassins regions, to the north-east by the Centre-Ouest region, to the east by the Republic of Ghana and to the south by the Republic of Côte d'Ivoire. The region has twenty-eight (28) communes, four (4) of which are urban and twenty-four (24) rural.

2.1.   Presentation of the municipality of Gbomblora

2.1.1. Location of the commune

First established as a department in 1983, then as a commune in 2006, the rural commune of Gbomblora is located in the province of Poni, in the South-West region of Burkina Faso. The main town, Gbomblora, is about 20 km from Gaoua, on the road leading to Batié. The commune has seventy-seven villages. The villages visited are shown in Figure 1.

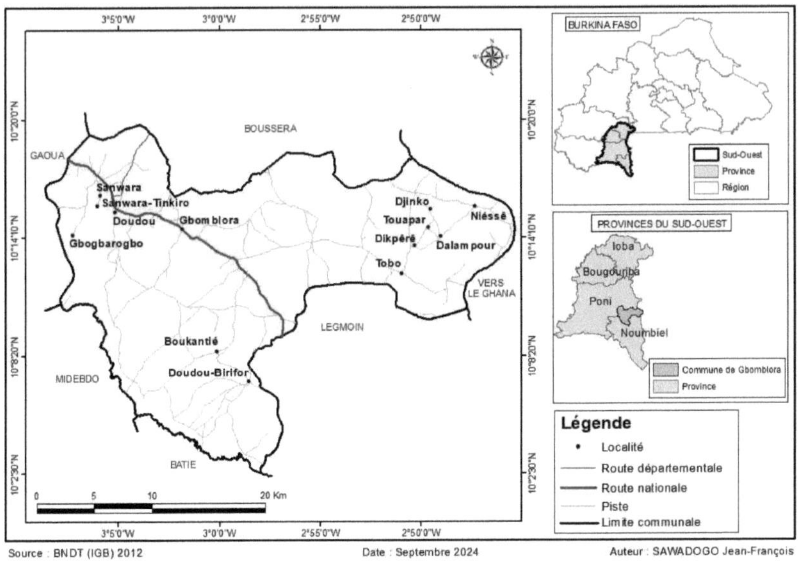

Figure 1 Location of the Rural Commune of Gbomblora

The commune is bordered to the north by Bousséra, to the west by Gaoua and Midebdo, to the south by Batié and Legmoin and to the east by the Republic of Ghana. The River Mouhoun forms the natural boundary between the two countries.

2.1.2. Physical environment

2.1.2.1. Climate

The south-western region of Burkina Faso is one of the wettest in the country. The data used to characterise the climate comes from the Gaoua synoptic station, which has a comparable climate. It is a Sudanese-type climate with two seasons, a rainy season and a dry season, each lasting almost six (6) months.

Figure 2 shows the interannual variability of rainfall in Gaoua. Although the general trend is upwards, there is considerable variability. Very wet years record total rainfall of more than 1,400 mm. These were recorded in 1992 and 2019. Other years with good rainfall were 1997-1998, 2000-2001, 2010-2011 and 2015-2016. Generally speaking, two successive years receive rainfall in excess of the trend line. These are followed by one or sometimes two years with less rainfall. The trend line shows peaks in rainfall, followed by periods of less rainfall. The lowest rainfall was in 2006.

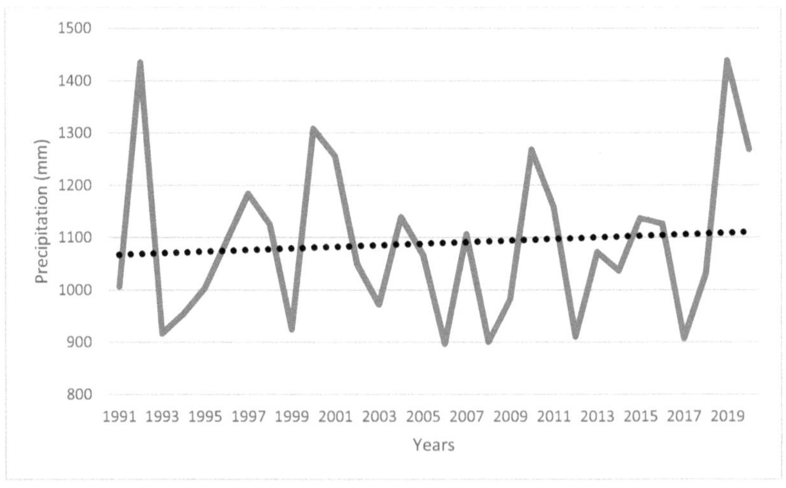

Figure 2 Normal rainfall trend 1991-2020, Gaoua synoptic station

The average annual rainfall for this normal is 1089.1 mm.

The annual rainfall seems to correspond well to the number of rainy days. However, between 2010 and 2013, a large number of days were recorded. From 2014 onwards, this number fell until 2019, when it rose sharply.

The trend in the number of rainy days is upwards, and greater than the increase in rainfall (Figure 3). The highest number of days was observed in 2019 (101 days) and the lowest in 1991 (74 days). The average number of rainy days for the normal period is 88.6 days.

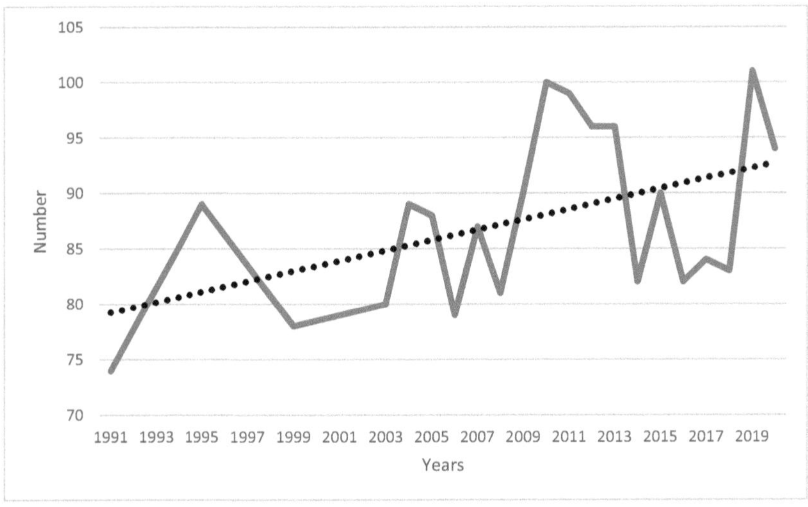

Figure 3 Number of rainy days per year of the 1991-2020 normal, Gaoua synoptic station

The average annual minimum temperature is 21.6°C and the maximum 34.2°C. This indicates a warm environment (Figure 4).

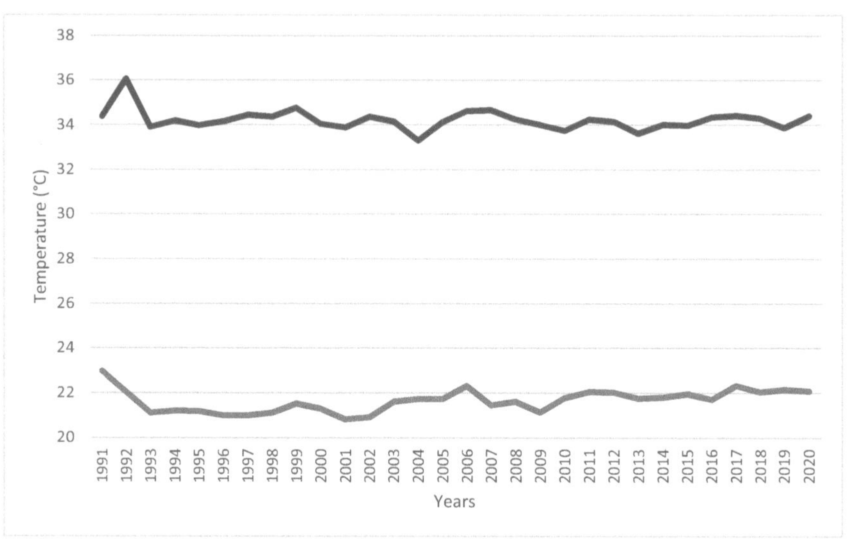

Figure 4 Maximum and minimum temperatures of the 1991-2020 normal, Gaoua synoptic station

Temperature amplitudes are consequently high, but are tending to fall significantly (Figure 5). 1991 was a special year in that it recorded the lowest amplitude. Since 1992, however, these amplitudes have been decreasing steadily, with significant variability from one year to the next.

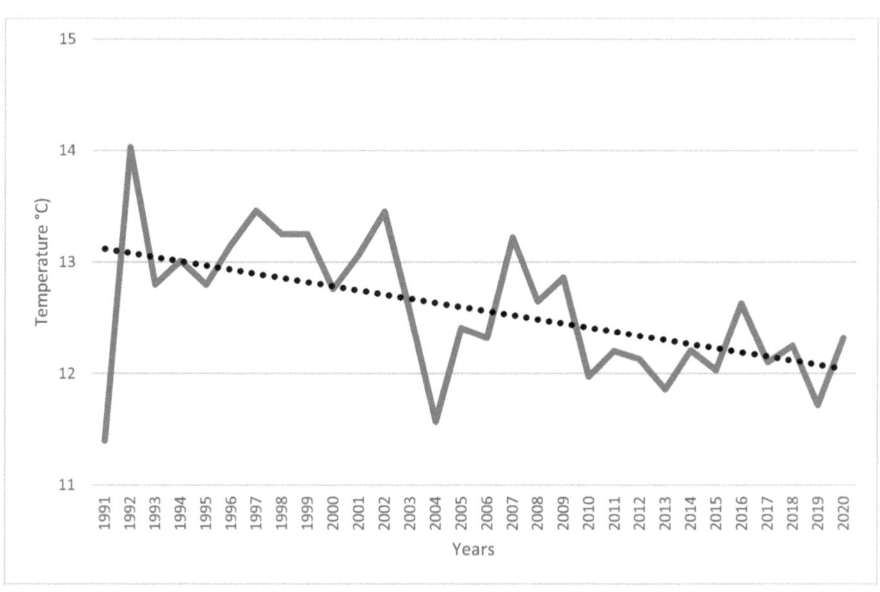

Figure 5 Annual thermal amplitudes of the 1991-2020 normal, Gaoua synoptic station

The average amplitude for normal is 12.6°C.

On the other hand, average annual wind speeds are trending upwards. This indicates that winds are becoming increasingly violent, with the potential to cause significant material and human damage. This upward trend is mainly due to the higher wind speeds recorded in 1994-1995, and also between 2008 and 2011. In the latter years, 2012 to 2014, speeds were low, but the averages for 2015 and 2016 were higher.

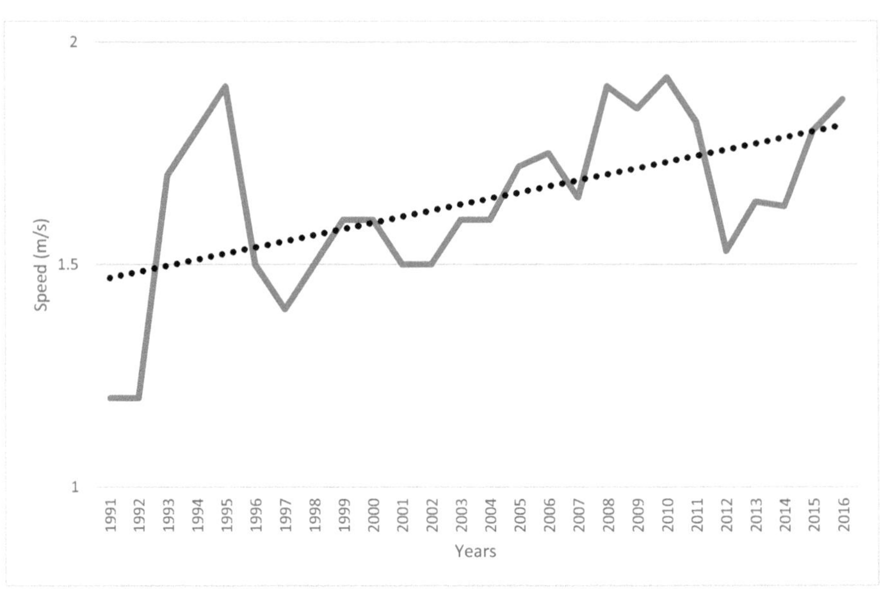

Figure 6 Mean annual wind speeds for the 1991-2020 normal, Gaoua synoptic station

The climate in Gaoua and the surrounding area is of the Sudanese type, characterised by two distinct dry and rainy seasons. The umbro-thermal diagram (Figure 7) shows that, on average, all the months received the same amount of water during the normal 1991-2020 period. These monthly averages conceal numerous irregularities.

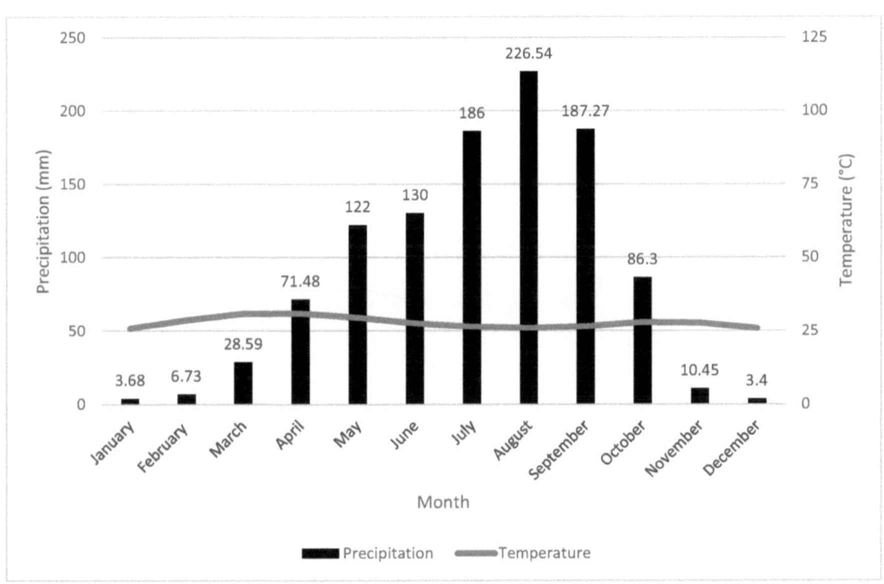

Figure 7 Umbro-thermal diagram for the 1991-2020 normal, Gaoua synoptic station

The rainy season begins in earnest when the average monthly rainfall is greater than or equal to the average monthly temperature. According to Figure 7, it runs from April to October, i.e. seven months of rainy season. The dry season lasts only five (5) months, from November to March. Rainfall moderates temperatures during the wet season. Temperatures peak sharply in March and again in November. December and January have low temperatures.

2.1.2.2. Hydrography

The hydrography is organised around the main river, the Mouhoun, which forms the natural boundary between the commune and Ghana. There are also intermittent rivers that provide water. There is a reservoir at Tobo, which is used by livestock farmers at the end of the rainy season. It dries up quickly. During the dry season, the Mouhoun has its low-water periods, but is the preferred watering place for animals.

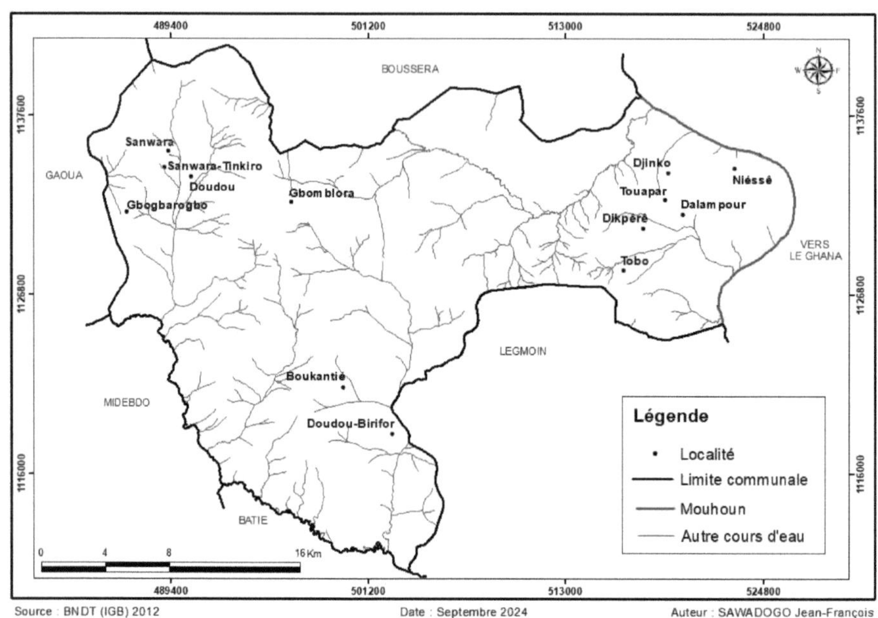

Figure 8 Hydrography of the commune of Gbomblora

2.1.2.3. *Relief and soils*

The terrain is rugged, especially in the western part. There are hills, the highest of which is Mont Koyo (592m), and rivers that flow into the Mouhoun. The eastern part consists more of plains and low-lying areas, some of which have been developed.

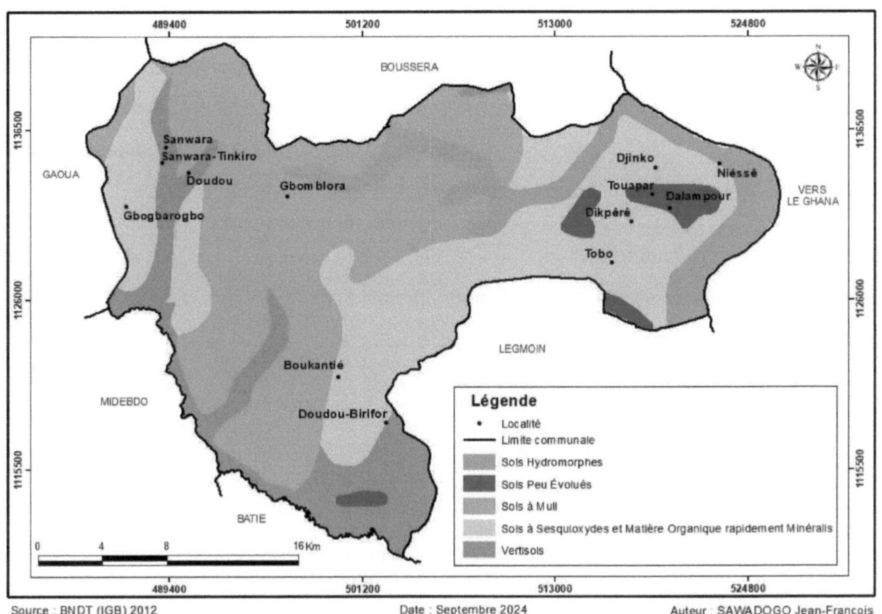

Figure 9 Soil in the rural commune of Gbomblora

There are five types of soil in the Gbomblora area. These are :

- Hydromorphic soils located along the River Mouhoun and one of its tributaries;

- Poorly developed soils at Dalampour and Touapar, west of Dikpéré and south of Tobo and Dou-Birifor ;

- Mull soils extending from Gbomblora to the south ;

- Soils with sesquioxides and rapidly mineralised organic matter from Djinko via Tobo, Boukantié and Doudou-Birifor ;

- Vertisols in the villages of Sanwara, Sanwara-Tintiro, Doudou and the borders with neighbouring communes to the east and Batié.

Almost all the soils are sandy clay or sandy clay. There are, however, some gravelly, silty or clay-loam soils in small proportions. These soils are rich and favourable to agriculture. Vegetation

The rapid run-off of rainwater means that the hills cannot support dense tree vegetation. Instead, there are only shrubby savannahs, where the herbaceous cover is extensive and used for livestock farming. The dominant plant species are : *Anona senegalensis, Isoberlina doka, Alzelia africaca, Diospiros mespiliformis, Lannea microcarpa, Acacia senegal, Kaya senegalensis, Detarium microcarpum, Cassia siberiana, Sclerocaria berrea*, etc.

The courses include flooded lowlands where rock forests develop. There are also savannah trees. The dominant species are *Vitex doniana, Raphia sudanica, Elaeis guinéensis, Bombax costatum, Ficus sp, Anona senegensis, Cassonia arborea*, etc.

2.1.2.4. Land use

The north-west and east of the commune of Gbomblora are the domain of shrub savannahs, no doubt due to greater human pressure. Rainfed crops and agroforestry occupy large areas here. Wooded savannahs and gallery forests occupy the western part of the commune, where human pressure appears to be lower.

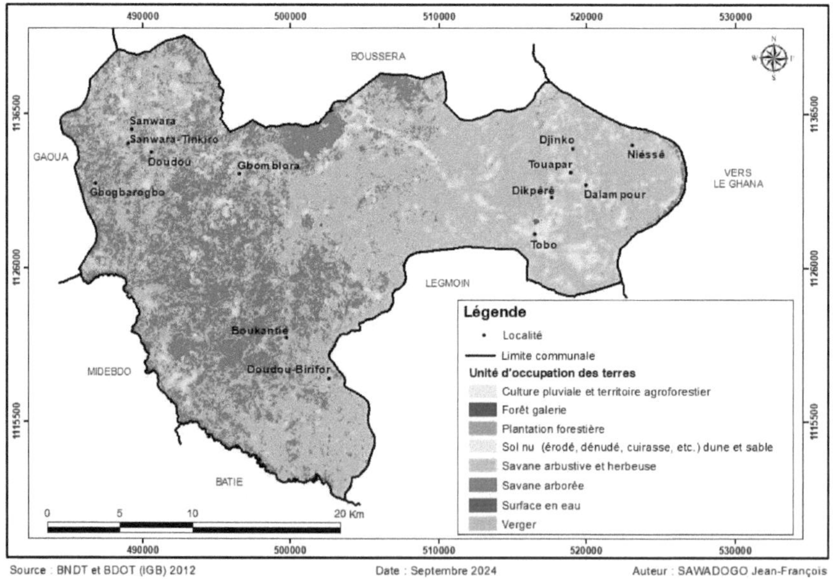

Figure 10 Land use

There are also orchards, particularly cashew orchards, and forest plantations.

2.1.3. Human environment

2.1.3.1.   Ethnic groups and population movements

The left and right banks of the Mouhoun river are inhabited by Lobis, Birifor and Dagara. Some villages are inhabited by sedentary agropastoralists. Other ethnic groups have been attracted by gold panning. The Burkinabè bank was populated by these peoples who came from Ghana in search of more fertile land. This migration continues to this day. For the purposes of customary rites, there are frequent movements to Ghana and Côte d'Ivoire.

These two countries receive large numbers of migrants, particularly during the dry season. Young people stay here to do seasonal work. They return to their homelands at the start of the rainy season.

Increasingly, the gold-panning sites, which are legion in the commune of Gbomblora, are attracting a particularly juvenile population. As a result, many children are dropping out of school, consuming illegal substances and engaging in prostitution.

2.1.3.2.   Population breakdown

In the 2006 general population and housing census (RGPH), the rural commune of Gbomblora had 18,793 inhabitants, 52.6% of whom were women. In 2019, there were 25,169, and the proportion of women had fallen sharply to 51.8%. Children under the age of 5 accounted for 14.3% of the total, those aged between 5 and 15 for 30.4%, those aged between 15 and 6 for 50.8% and those aged 6(and over for only 4.5%.

The population density was 44 inhabitants per km2. Some of the villages listed are farming hamlets with only a few inhabitants. Most of the population is concentrated in the largest villages, including the main town.

2.1.3.3.   Religions practised

Animism is widely practised, including through various initiations. The passage to adulthood is marked by rites called Boor for the Lobi and Booré for the Dagara. It is organised every year and involves young people of at least twelve (12) years of age. Participants are recruited from their communities in Burkina Faso and Ghana. Another ceremony organised every seven (7) years is the Djoro. It takes twelve (12) months to prepare and three (3) months to implement.

In addition to these traditional religious practices, revealed relationships are practised by minority populations. Christianity seems to be well established among the Dagara. Islam is practised by all settled agropastoralists.

2.1.3.4.   *Administrative and social organisation*

A prefect represented central government. With the advent of coups d'état, communes are now managed by special delegations presided over by prefects. Prefects act as chairmen of the special delegations, instead of elected municipal councils. Each recognised village is represented by a delegate and his or her deputy.

Each village also has a village chief who manages the fetishes and a land chief who is responsible for land management. It is the latter who carries out all the rites and sacrifices relating to land. Each lineage has its own land, managed by the men who have the right to use it. Women have access to the land through their husbands. Young people work the land under the guise of their fathers until the latter decide to give them full rights. They then form separate households and depend little on their fathers. Land is acquired through inheritance. It is also done by gifts to those who ask for them or by loans over specific periods of time. Acquisition by sale did not yet exist.

2.1.4. Socio-economic activities

Socio-economic activities are dominated by the primary sector and some mining activities.

2.1.4.1.   *Agriculture*

Food crops are the dominant form of agriculture. It is practised primarily to feed the family and secondarily for the market. Sorghum, the locality's leading crop, accounted for 56.43% of sown area, maize for 26.88%, millet for 13.08% and rice for 0.36%. Rice is grown in flooded lowlands, some of which have been developed. But yields are low overall, except for rice and maize. Yam, sweet potato, voandzou and cowpea are also grown, most of which is consumed by the family. The rest is sold in the markets.

Alongside this subsistence farming, cash crops are also grown. The main cash crops grown in the commune are groundnuts, soya and cotton. These crops account for 91.33%, 3.77% and 1.88% of the area respectively. This more intensive cash crop benefits from greater investment. Food crops, on the other hand, are more extensive, and their growth is due to the increase in the area under cultivation.

2.1.4.2.   Breeding

Livestock farming is also extensive. It is made up of sheep, cattle, goats, pigs and poultry. There are still some hardy local breeds, but they are better adapted to the area's very rainy climate, where contagious bovine pleuropneumonia, anthrax, trypanosomiasis and parasites are rife. The more productive breeds brought in by settled pastoralists are more exposed.

The departmental service does not have enough staff or adequate logistical resources to cover the commune, which has no specific livestock markets.

2.1.4.3.   Forestry and silviculture

The hills are privileged places where sacred village forests are protected. Some of these sacred forests are off-limits to foreigners, who will be punished as a deterrent. Even the animals of agropastoralists do not go there. Natural resources are thus preserved.

The commune also has large areas of fallow land, particularly in these hill forests, which are reserves. As a result, the local people gather and process non-timber forest products. Traditional and modern beehives are set up to collect honey. Game drives are organised to kill wild animals such as hares, ourébis, duikers and buffalo. The Gbomblora and Tobo reservoirs are used for fish farming and traditional fishing activities.

2.1.4.4.   Trade

There are only two markets in the commune, at Doudou and Tobo. Not even the main town has one. These markets are frequented by traders from Gaoua, mainly from Doudou. Inhabitants of surrounding villages come to sell agricultural produce and buy manufactured goods. The Tobo market is also frequented by traders from Ghana, but also from the neighbouring municipality of Legmoin and even from Batié, where the markets are better attended.

2.2. Study methodology

The following methodological stages were observed: document review, direct observations in the field and interviews with resource persons.

2.2.1. Documentary review

Internet searches did not yield a large number of documents, particularly relating to the start of the agricultural campaign in this rural commune. He provided us with the document of the 2015-2019 communal development plan, which had expired. A doctoral thesis on geography in the rural commune of Nako, in the province of Poni and the South-West Region was also used. Statistics, particularly on the population, are provided by a file widely distributed by the National Institute of Statistics and Demography. These are data from the latest general population and housing census for 2019.

2.2.2. Field observations

A field mission was carried out in the commune in July 2024. It visited the villages of Doudou, Sanwara, Sanwara-Tintiro, Gbogbarogbo, Gbomblora, Tobo, Dalampour, Dikpéré, Djinko, Niessè, Touapar, Doudou-Birifor and BouKantié. Photographs were taken to document these field observations.

2.2.3. Interviews with resource persons

During the mission, observations that raised questions were addressed to the agents specialising in the environment, animal resources, agriculture and the commune's land department. The questionnaire was not very structured and was designed to provide answers to specific questions.

3. Results obtained

This mainly concerns the dates on which the rainy season starts and the practices and techniques used to adapt and make the most of the programmes implemented.

3.1. Optimum sowing dates

Figure 10 shows three main trends. The first trend from 1991 to 1997, when the optimum sowing dates were between mid-April and the first dekad of May. The second period, from 1998 to 2008, saw marked variability in the start dates of the season. Optimum dates were as early as the beginning of April and some as late as 25 May. Very early starts were observed in 1998 on April 3 and late starts in 2002 on May 27. The variability became more pronounced in 2008, with a very early start on 3 April. The onsets became later and later, with a record in 2019 on 17 June. Late starts are increasingly frequent, between the end of May and mid-June, during this second trend. The general trend is upwards, and farmers should expect high variability and increasingly late optimum sowing dates in the coming years.

Figure 10 shows the later start of the rainy season. It reflects the ideal start to the season when sowing can be carried out under good weather conditions.

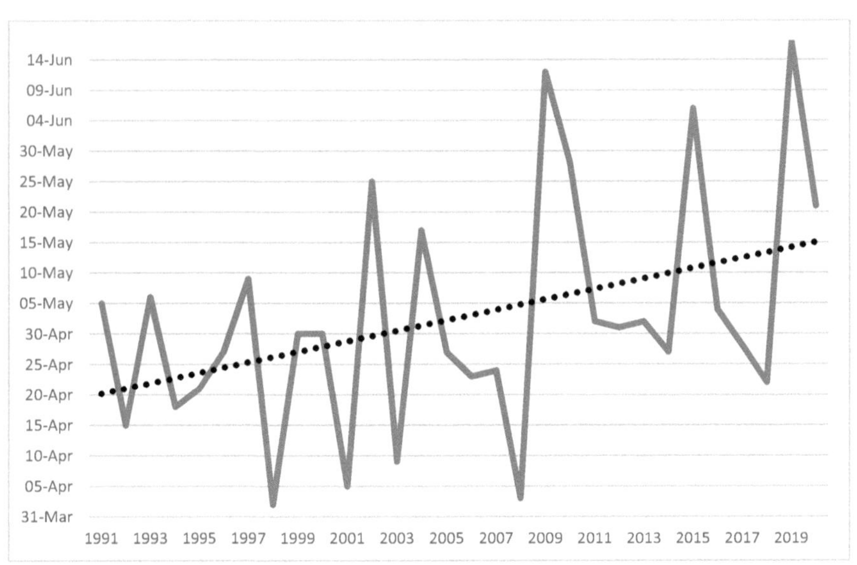

Figure 11 Optimal dates for the start of the rainy season, Gaoua synoptic station

3.2. Rainfall at the start of the rainy season in 2024

Table 1 below shows the rainfall recorded in Gaoua from January to July 2004.

Table 1 Daily rainfall at Gaoua station, Poni province, South-West region

	January	February	March	April	May	June	July
1	0,0	0,0	0,0	0,0	0,0	0,0	0,0
2	0,0	0,0	0,0	0,0	0,0	0,0	0,0
3	0,0	0,0	0,0	0,0	0,0	19,5	20,1
4	0,0	0,0	0,0	0,0	0,0	2,4	0,7
5	0,0	0,0	0,0	0,0	9,6	0,0	0,0
6	0,0	0,0	0,0	0,0	2,6	0,0	2,6
7	0,0	0,0	0,0	0,0	0,0	0,0	5,3
8	0,0	0,0	0,0	2,5	0,0	0,0	2,2
9	0,0	0,0	0,0	0,0	0,0	35,6	16,8
10	0,0	0,0	0,0	0,0	0,0	36,8	0,0
11	0,0	0,0	0,0	0,0	0,0	0,0	0,0
12	0,0	0,0	0,0	0,0	9,2	5,0	23,5
13	0,0	0,0	0,0	0,0	5,4	0,0	0,0
14	0,0	0,0	0,0	0,0	0,0	0,0	0,0
15	0,0	0,0	0,0	0,0	0,0	0,0	0,0
16	0,0	0,0	0,0	0,0	7,4	5,2	0,0
17	0,0	0,0	0,0	0,0	0,0	0,0	36,4
18	0,0	0,0	0,0	0,0	0,0	0,0	0,0
19	0,0	0,0	0,0	0,0	0,0	2,3	0,0
20	0,0	0,0	0,0	0,0	0,0	0,0	0,6
21	0,0	0,0	0,0	0,0	0,0	2,3	0,0
22	0,0	0,0	0,0	0,0	0,0	0,0	0,0
23	0,0	0,0	0,0	0,0	0,0	0,0	0,3
24	0,0	0,0	0,0	0,8	0,0	0,0	0,0
25	0,0	0,0	0,0	0,0	0,0	0,0	0,0
26	0,0	0,0	0,0	0,0	0,0	0,0	0,0
27	0,0	0,0	0,0	27,8	0,0	0,0	0,7
28	0,0	0,0	0,0	0,0	0,0	0,0	4,7
29	0,0		0,0	0,0	9,3	22,7	27,3
30	0,0		0,0	0,0	0,0	0,0	0,0
31	0,0		0,0		0,0		0,0

For the year 2024, the first three months of January, February and April received no rain. An initial rainfall of 2.5 mm fell on April 8, followed by a more substantial 27.8 mm on April 27. Seven (7) days passed and on May 5 and May, 9.6 and 2.6 mm of water fell successively.

Throughout May, the rainfall recorded was barely 10 mm. This light rain was more favourable to weeds. In early June, there were three (3) heavy rains, two of which fell in succession on the 9th and 10th, and another on the 25th. During July, successive rains became more frequent, from the 6th to the 9th and on the 29th, 30th and 31st July.

The number of rainy days rose from zero (0) in January to March, to 3, 6, 9 and 12 in April, May, June and July respectively.

Monthly rainfall totals are shown in Figure 12 below.

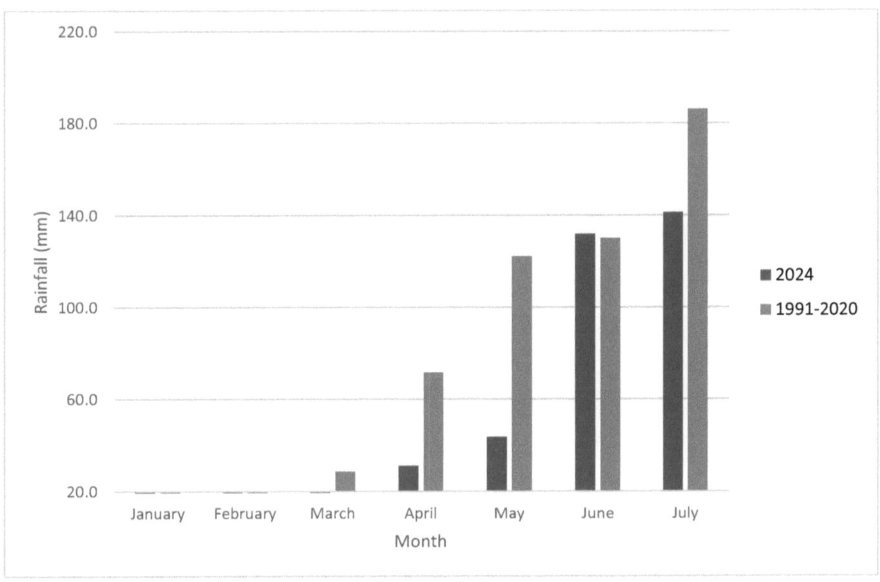

Figure 12 Monthly rainfall totals in Gaoua in 2024 and the 1991-2020 normal

31.1 mm of rain fell in April 2024, compared with 43.5 mm in May, 131.8 mm in June and 141.2 mm in July. Compared with the 1991-2020 normal, the months of January and February received rainfall close to zero (0). With the exception of June, which received less total rainfall, the months of March to May showed much greater deficits than July. As a result, there was a rainfall deficit in 2024.

The peak will certainly be reached in August before the end of the season in September. What techniques and practices are used by farmers in the commune of Gbomblora?

3.3.   Practices and techniques observed

These observations were made in July 2024. It can be assumed that the optimum sowing date is well past at this time. The practices and techniques observed are described below

3.3.1. Weed control

As mentioned above, any month can receive rain. Most of this rain falls out of season and is often referred to as mango rain. This is because it falls just before the early mangoes ripen, helping them to ripen properly. Around the concessions, the ground is covered by grass, which is often tall. This explains the use of herbicides.

Photo 1 Tall grass invading hut fields

3.3.1.1. *Use of herbicides and/or pesticides*

According to the officer in charge of agriculture, these are chemicals, most of which are on the approved list published by the Ministry of Agriculture. However, the proximity of the Ghana border allows other prohibited products to be smuggled in. But he deplores the adaptations made by some of them to the use of approved products.

Photo 2 Two farmers spraying a herbicide without any protection

In the field, unprotected farmers are spreading herbicides. They naturally inhale it, exposing themselves to disease.

To dry out the herbs more quickly, some farmers exceed the prescribed doses. And to ensure that the dose used will be effective in drying out the weeds quickly, some farmers go so far as to test the dose on the tongue. The result in the field is total or partial drying out of the weeds in the sprayed areas.

Photo 3 Dried grass in a field

These dried-up weeds are buried after ploughing or when mounds are made. These practices and techniques are used primarily in the hutment fields, where the first rains and domestic waste are favourable to the development of weeds. In more remote fields, preparatory work is carried out.

3.3.1.2. *The preparatory work*

Several situations were encountered: land clearance, fields with minimal preparation, ploughing and mounding.

- Land clearing

In the rural commune of Gbomblora, there are reserves of land, including old fallow land, which are being cleared to be turned back into fields. Utilitarian species such as *Vitelaria paradoxum, Lannea microcarpa and Detarium sp.* are preserved. The others are cut so that many of them do not die. Trunks can be seen scattered across the field. These trees grow back during the dry period and need to be cleared before sowing, to prevent the shade from reducing crop yields. These are agroforestry parks where the trees enrich the soil and protect it from water erosion. Their roots continue to protect the soil.

Photo 4 Clearing of former fallow land

The land on these former fallow lands had been naturally reconstituted. These are rich soils where seedlings are sown without any further development. Weeds also thrive here, competing with crops for water and nutrients. Ploughing and other weeding operations eliminate them in large numbers.

- Old undeveloped fields

In many situations, seed is sown directly without any preparation. This is done on old fields and new clearings.

Photo 5 Sowing in a lowland field with little management

In this lowland field, only manure was spread just before sowing. This field supports several crops. There are cowpea plants and maize further on. The low-lying position of the field makes it vulnerable to flooding. The closer you get to the river bed, the more water-demanding species that can withstand short-term flooding are sown, such as maize. This is the case with maize. Species that are more vulnerable to flooding, such as cowpea, are sown higher up on the major river bed.

- Staggered or in-line mounds

In photo 5, mounds have been planted in staggered rows. Because of the high rainfall, furrows were created between the mounds to drain off rainwater. This practice prevents the crops from suffocating, especially during exceptional rainfall, and results in poor yields. Seedlings are also sown at the top of the mound. Flooding is only possible when the whole field is submerged in water, during exceptionally heavy rains.

Photo 6 Mounds supporting several crops

Mounding is hard work. It is easy to see that part of the field has no mounds. Mounds are made in the traditional way, using rudimentary tools such as the daba.

These staggered mounds also support yam fields. Dry wood is planted on these mounds, which the yam plants use for climbing. The mound allows the yam pod to grow. When these pods become large, the mounds crack.

Mounds are also made in rows (Photo 6). They are considered parallel and perpendicular to the slope to slow run-off and soil erosion. Furrows are also left between them to allow water to drain away. The in-line mounds are also separated by larger furrows to collect water from heavy rainfall. In the event of exceptional rainfall, these furrows encourage more run-off and erosion.

The cropping systems on the hillocks are associative. Cowpeas, sorghum and maize are grown at the top of the mounds.

Photo 7 Linear mounds supporting a crop association

- Adapting to the staggered onset of the rainy season

Farmers are aware of the unpredictability of rainfall and, based on the observation of natural indicators, are taking the risk of sowing early, as soon as the first rains are deemed useful. During our visit in the second half of July, we came across fields of ripening maize. The vast majority were located in low-lying areas that collect run-off water and where the soil is rich.

In almost all cases, the fields are in a state of lifting, as shown in photos 5 to 7. There are two different situations. Near the Mouhoun river, where the gradients are lower and the fields more numerous and occupying large areas, the hut fields are almost all sown. The fields further away, particularly the clearings, continued to be sown.

In the eastern part, the more distant fields were sown. The hut fields in all the localities visited were invaded by weeds. Farmers use herbicides to dry them out before ploughing. During the visit, these fields had not yet been sown.

- Arboriculture

The tour included a number of cashew orchards. According to the officer in charge of the environment, these orchards have been planted in the traditional way, with no support from his department. As a result, the cashew trees are not the right species and yield very little. Spacing and a number of recommendations have not been taken into account.

Photo 8 Cashew orchard

The owners do not make the most of them. These unprofitable orchards are not maintained. This is due, among other things, to weaknesses in agricultural management, in the

4. Constraints encountered

These field observations lay bare the constraints faced by this type of agriculture. These include weaknesses in agricultural supervision and producers' capacities, the narrowness of markets and rural finance, and competition from gold panning.

4.1. Poor agricultural supervision

In July 2024, the government technical services present in Gbomblora were departmental services:

- Environment, water and sanitation: these are mainly forestry officers who assist local people in exploiting natural resources such as timber and non-timber forest products, including fishing and hunting resources. They are also feared for the fines imposed on fraudsters. Hunting and gathering products are systematically hidden to avoid being summoned to the office. The equipment used can also be seized.

- Agriculture, animal resources and fisheries: one animal resources officer was available, more concerned with inspecting meat in the two markets. Another agricultural officer covers the rural commune of Gbomblora, but is assigned to the commune of Gaoua, where he lives. They are responsible, each in their own area, for supervising agricultural producers so that they can improve their production systems.

- Special delegation resulting from the dissolution of the municipal council: the agent in charge of the rural land service (SFR) to secure the delivery of certificates of land ownership at the request of families.

These technical officers had to travel around the land and sometimes give technical advice on the questions asked. But only the environment and animal resources officers had service motorbikes in fairly good condition. The agricultural officer came with his own motorbike, which is not very suitable for this area with its bad roads and steep gradients. The SFR agent used the town hall's motorbike, which was used for other tasks.

They do not have enough fuel to enable them to visit all the villages and build producers' capacities. For this reason, their work depended on the funding obtained by the projects and the civil society organisations that requested their support. As a result, they have very little involvement with farmers.

This situation is exacerbated by the poor state of the roads and the inaccessibility of some villages (Compaore et al., 2023; Pouliot et al., 2016). Because of heavy rain, a first appointment could not be honoured. The water had taken up a large part of the main riverbed and it was necessary to cross the river and walk to the villages of Doudou-Birifor and Boukantié. At the second meeting, another heavy rain fell while we were already in the village. The villagers helped us by pointing out parts further upstream so that we could cross the stream more easily.

The agents are insufficient in number and have means of transport, but lack the fuel and financial resources to travel to the villages and properly supervise the farmers. The poor state of the roads is also a major obstacle to supervision. This weakness in the supervision of producers is compounded by the narrowness of markets.

4.2.   Tight markets and local finances

There are only two major markets in the rural commune of Gbomblora. The commune's main town has none. The first market is in Doudou, less than 20 km from Gaoua, the provincial capital of Poni and the South-West region. It is served by the national road that links Gaoua to Kpéré and on to Côte d'Ivoire. When we visited, the road was being paved. There were many diversions, made to build the crossing structures. Many traders, mainly from Gaoua, come here to sell manufactured goods and buy agricultural produce for resale in the town of Gaoua.

Further south, after the main town, a fork leads to the Tobo market. This road is passable in all seasons, but has a number of bumps and speed bumps that slow down traffic. These two internal markets are frequented by farmers who come here to sell their produce and buy products from the larger markets in Gaoua, Legmoin and Batié.

To develop their agricultural production activities, farmers do not yet have a microfinance institution in the commune. This means that access to credit is very poor. This is reflected in the low level of agricultural intensification. Most farming equipment is still rudimentary. Income from the sale of agricultural produce hardly allows for the purchase of more modern means of production. Despite good soil fertility, agricultural yields remain low overall. This type of farming also uses few agricultural inputs. Few fields are enriched with mineral fertiliser. Livestock produce waste that is spread on the fields to improve their fertility and achieve much higher yields.

The mobilisation of local savings is also minimal. The few farmers who make financial transactions are forced to go to the major centres in Gaoua, where there are banking and microfinance institutions.

Farmers in the rural commune of Gbomblora remain poor on the whole. This poverty is even more acute in villages that are difficult to reach. The whole village of Boukantié, according to the inhabitants, had only a few motorbikes, including that of the village delegate. All the villagers therefore travel on foot. However, gold panning is widespread in the western part of the commune.

4.3. Predominant role of gold panning

One of the most striking features of the landscape of the rural commune of Gbomblora is the existence of gold panning, as evidenced by the many holes left after mining. The entire hilly western part of the commune is home to numerous gold panning sites. During the rainy season, these sites are closed to avoid the risk of accidents and loss of human life (CILSS, 2012; MJE, 2007; Palé, 2020).

These hills are dotted with gold panning holes, which are rather waterlogged and unexploited. These holes are often polluted by the uncontrolled use of cyanide (Burkina Faso, 2013; UNEP, 2008; Yaméogo, 2015). This is why they constitute vast fields unsuitable for agriculture.

Gold panning is a more profitable activity, attracting people from all over Burkina and even from the West African sub-region. Brave young people give up farming in the hope of a better future. This activity also leads to people dropping out of school. At some of the most important sites, economic activities such as the sale of drinks, often prohibited, drugs, etc. are created. These places are also conducive to prostitution.

Photo 9 Holes left by gold panning

Gold mining is robbing agriculture of its able-bodied workforce. The influx of people from elsewhere increases the demand for basic necessities. As a result, prices are rising, and many people are struggling to afford three meals a day on the gold-mining sites. In Batié, as one pork vendor put it, everything is expensive. This was confirmed by an environmental officer, who said that roosters were sold for around six thousand (6,000) CFA francs. But what role does agriculture play in local development planning?

4.4. Agriculture in local development planning

In 2024, the project for sustainable management of communal landscapes for REDD+ chose to formulate an integrated communal development plan for REDD+ (PDIC/REDD+) for the commune of Gbomblora. It is largely focused on reducing emissions from degradation and deforestation. Carbon sequestration and, to a lesser extent, adaptation are the objectives sought. Three conservation areas have been chosen by the local people, who have undertaken not to carry out any agricultural activities there. These areas will be protected by hedgerows or firebreaks and enriched by the planting of local and useful plant species. Additional activities, including the construction of various facilities such as vaccination parks, cattle tracks, market gardens, rice-growing areas, etc., have been initiated to minimise human pressure on these conservation areas and ensure their sustainability. Capacity-building activities are aimed at improving the management of these infrastructures.

Strictly agricultural activities are set aside to make way for carbon sequestration and the management of environmental and social impacts. Over time, the quantity of greenhouse gases sequestered can be quantified and sold on carbon markets. This will generate additional income for local communities.

This PDIC/REDD+ will be implemented with initial funding of two years and additional funding of one year. At the outset, a municipal development plan for 2015-2019 was implemented. This plan also focused on physical achievements, relegating technical and managerial strengthening activities to second place. Objective 6 aimed to increase cereal and market garden production, with water reservoirs and facilities to be built. Objective 7 also aimed to increase livestock production by creating pastoral areas and vaccination parks.

The aim of these projects was to provide communities with means of production that are not available to farms and farming households. However, farming is family-based and there is no such thing as a community farm. The market economy quickly broke up the few farms that existed. Households were set up, followed by the allocation of land for farming. However, all the local development planning in Gbomblora was based on the community's achievements. This satisfied the needs of the politicians, especially the political advisers. They can boast that such achievements were made under his watch. A way for this politician to attract the sympathy of voters and their votes in local elections.

4.5. The South-West region in agricultural development programmes

Two programme directories were used, those for 2020 and 2023.

4.5.1. Directory of 2020 programmes

A number of nationwide programmes are underway in the South West region. The vast majority of these projects involve construction work. They involve water development, mechanisation through the acquisition of agricultural equipment and infrastructure. Some of them have focused on developing sectors, securing land tenure and managing related conflicts, and building the capacity of both farmers and management.

The nationally-funded Agricultural Production Intensification Programme provides farmers with improved seeds to improve yields and ensure food security. It covers both rainfed and off-season crops. One of the programme's components strengthens management capacity by paying agents.

It is through such programmes that motorbikes are acquired and made available to the agents of the technical support units to facilitate their travel and supervision in the field. The regions that benefit from the intervention of several programmes are the most privileged, and their agents have the opportunity to have their means of transport quickly renewed.

Apart from national programmes, in 2020 the South-West benefited from only three programmes: the Irrigation Programme in the Great West (PIGO), which was implemented in all four (4) provinces.

4.5.1.1.   Irrigation programme in the Grand-Ouest

On the main road from Gaoua to Batié, low-lying areas are used for rice growing. On the road leading to Dalampour, some of these facilities have deteriorated. The aim is to improve food security and increase people's agricultural incomes. The DPA plans to carry out a number of projects, including development work, the construction of warehouses and processing units, threshing floors and crossing structures. The expected results are

- 2,000 ha of lowland rice-growing areas have been developed and put to good use;

- 40 ha of market garden areas have been developed and put to good use;

- 80 warehouses for the storage and conservation of agricultural products have been built and equipped;

- 5 agricultural product processing units have been built and developed;

- 40 threshing-drying areas/ tarpaulins built and used ;

- Crossing structures have been built to open up certain areas with developed lowlands;

- at least 10,000 industry players are trained in a range of subjects (management of facilities, production techniques, financial support, etc.).

These achievements are very useful because they provide agricultural producers with means of production that are beyond their means.

4.5.1.2. *Agricultural Development Programme*

The Agricultural Development Programme, implemented by GIZ, the German development cooperation agency. The South-West is one of the preferred regions for intervention by GIZ, which has set up a regional office in Gaoua. The PDA aims to improve the farm results of very small, small and medium-sized processing and marketing companies in the rice and cassava sectors. The expected results are

- The framework conditions for improving the performance of farms and VSE-SMEs in the rice and cassava sectors have been improved;

- The farming capacities of very small businesses are strengthened to enable them to implement their business model;

- the range of services on offer has been expanded in the rice and cassava sectors.

Food crops, which account for almost all cultivated land, do not fall within the scope of the PDA.

4.5.1.3. *Support Programme for Agricultural Sectors in the South-West, Hauts-Bassins, Cascades and Boucle du Mouhoun Regions*

The aim of the programme was to achieve sustainable improvements in food security and incomes for farmers involved in the production and marketing of products in the rice, market gardening, sesame and cowpea sectors. It aimed to achieve the following results

- Access to inputs, equipment and support/advice is promoted;

- 3 000 ha of lowlands are developed or rehabilitated;

- 500 ha of small market garden areas have been developed;

- 300 ha of market garden developments using water-saving irrigation technologies have been completed;

- 100 km of site access tracks have been built

Infrastructure development also plays a key role in this programme. It includes rural tracks to improve mobility and access to markets for trading activities.

In these three projects, which include the South-West region, cash crops such as rice and manioc, as well as market garden crops, are being promoted. The physical works planned are favourable factors for the expansion of these cash crops. Urban development is an important market for these crops.

4.5.2. Directory of 2023 projects

In 2023, the security situation has led to the development of an action plan for stability and development. All programmes must be aligned with the plan's objectives. A new directory has been built. What is the position of the South-West region in this new plan, which complements the second national economic and social development plan?

4.5.2.1. *Support Programme for Agricultural Sectors in the South-West, Hauts-Bassins, Cascades and Boucle du Mouhoun Regions*

The programme continued to be implemented with the same objectives and expected results.

4.5.2.2. *Project to improve livestock mobility and agropastoralists' incomes through the use of mobile telephony and satellite imagery. Pastoral infrastructure component*

The overall objective is to improve the productivity, income, resilience and food security of pastoral and agro-pastoral populations in a context of climate change and security crises. The main expected results are

- By 2023, access to advice, inputs and financial resources will have increased by 15% for 450,000 livestock breeders, pastoralists and farmers in the intervention regions. As a result of the reliable and timely information shared via the mobility support service, advice on agro-pastoral operations, the ease of ordering inputs and access to suitable financial products (savings and loans), pastoralists and farmers will be in a better position to make informed decisions and invest in the productive resources of agro-pastoral operations.

- Between now and 2023, incomes and animal and agricultural productivity will be improved by 10% for 450,000 livestock breeders, pastoralists and farmers in the intervention regions. Thanks to better access to information, advice, inputs and markets, animal and agricultural productivity will increase, as will incomes. Achieving this result will contribute directly to the multi-annual plan of the Embassy of the Kingdom of the Netherlands, particularly to the indicators relating to food and nutritional security.

- By 2023, the GARBAL solution's business model is viable so that financial equilibrium is achieved. Its institutional viability is defined within the framework of a public-private partnership agreement.

This programme ended on 31 December 2023.

4.5.2.3.   *Project to improve the nutritional situation through agriculture*

This project aims to improve the food and nutrition situation through behavioural change in the target regions, including the South-West. The expected results are :

- three (03) studies linked to market-oriented agriculture, nutrition and school canteens are being carried out;

- the SHEP approach is implemented (200 MAAHM STD agents are trained in the SHEP approach, 1900 producers are trained in the SHEP approach, 57 agricultural equipment kits are made available to producers trained in the SHEP approach, 38 professional agricultural organisations are registered, 19 educational plots are set up);

- Activities to improve nutrition at community level are being stepped up (3,360 kg of nutritional inputs are being made available to CSPSs, 850 women are receiving small ruminant nuclei, 950 women are being trained in processing/preservation and storage techniques, 19 processing kits are being made available to women, 950 women are being trained in good food and nutrition practices, 800 people from households are being trained in WASH, 36 GASPAs and 19 husbands' schools are being set up);

- activities linked to school canteens are being stepped up (19 school gardens are being supported, 19 kitchens, 19 shops, 19 toilets and 19 boreholes are being rehabilitated, 171 education staff and 38 canteen workers are being trained in WASH, 420 pupils are being trained in techniques for producing organic manure, 19 schools are being equipped with kitchen and gardening kits and hand-washing facilities, 640 high-nutritional-value plants are being made available to school gardens).

4.5.2.4. *Project 2 of the programme to strengthen resilience against food and nutritional insecurity in the Sahel*

The overall objective is to improve living conditions and food and nutritional security for people in the Sahel. The expected results are

- Increase in agricultural production (plant/animal/fisheries) from (83,000/68,000/15,000) to (135,000/86,400/20,000) ;

- Increase in income per capita (man/woman): (GNI/capita) from 906 US dollars to 945 US dollars;

- Number of jobs created in the project area (% women) at 60% ;

- Number of beneficiaries (producers/breeders/fishermen who have adopted gender-sensitive climate change resilience practices) (% women) to 80,000 with 50% women.

4.5.2.5. *Small-scale irrigation project in the Grand Ouest and Eastern regions*

In a second phase, the irrigation project in the Great West was extended to the Eastern region. Its overall objective is to improve the living conditions and resilience of rural populations by strengthening their food security through increased production and income, in a preserved environment. The objectives are to

- Lowland infrastructure, market garden/irrigated areas and access roads are being built in a natural environment that has been preserved for efficient and sustainable use;

- Infrastructure to supply inputs, process and store agricultural produce, add value to products and provide access to markets has been put in place;

- The infrastructure built/rehabilitated is used on a long-term basis and maintained by the beneficiaries and their organisations with the support of local authorities and technical services;

- Small-scale producers manage water properly and intensify production in lowlands and market garden/irrigated areas;

- The beneficiary households process and market part of their rice harvest.

4.5.2.6.   Agricultural Development Programme, 2022-2025 module

The overall objective is to achieve food security and reduce poverty among the population through the intensification, processing and marketing of agricultural production in the programme area.

- the adoption of at least four (04) out of ten (10) recommended practices for agro-ecological and environmentally friendly production by 10 333 farms (3 875 run by women, 3 100 by young people) and 1033 processing enterprises (930 run by women, 310 by young people) in the cassava, rice and soya sectors;

- A 20% improvement in yield per hectare on the farms of the 15,500 supported producers, 25% of whom are managed by women and 20% by young people, compared with the yield of unsupported producers;

- a 20% improvement in the average incomes of 17,050 agricultural producers supported and 1,550 processing SMEs;

- the creation of 630 additional permanent jobs in 1,550 processing SMEs in the CVAs supported, including 350 for women and 150 for young people.

- the adoption of 8 out of 20 recommended food and hygiene practices by 1,800 people out of 3,000 trained, 80% of whom were women.

The results of this programme are disaggregated to take gender into account.

The number of projects in the South-West region rose from three (3) to six (6). Two (2) of these three (3) projects were renewed. In practice, these projects rarely cover the entire region. The results obtained do not make it possible to say which communes are covered by each of these projects.

But all projects want to have a significant and visible impact. That's why they choose to work in smaller areas.

The issues addressed concern food and nutritional safety. The sectors attract the most attention. Constraints at the start of the agricultural season and food crops are not specifically mentioned. As a result, the projects and programmes implemented provide few responses. As already mentioned, local development planning also fails to address them. In the case of the villages visited in the Commune of Gbomblora, agricultural producers seem to be left to their own devices.

Conclusion

Several communal plans have been implemented after the village management plans. But local development is clearly slow to really take off. And yet the agricultural potential of this commune of Gbomblora is real. This is the part of Burkina Faso where rainfall is spread over six (6) years or more, with an average of over 1000 mm per year. The soils are rich, especially in the western part. A student, who accompanied us to identify the sites where the projects would be carried out, said that the soils in Sanwara-Tinkiro are very rich and do not need any special amendment to produce high yields. But farmers lack the resources and modern know-how to exploit this high potential.

Farmers first use their local know-how, including staggered ridges, traditional clearing and staggered sowing, to ensure that agricultural production is focused primarily on meeting the food needs of the farm or household (Somda et al., 2014). The first hypothesis is thus confirmed.

For this to happen, local development planning needs to place family farming in the place of the community (Camilla Toulmin, 2003; Dugu et al., 2012; IED, 2014). This will enable the municipality and its councillors to acquire microfinance institutions and banks capable of training them in technical itineraries incorporating their local needs and know-how. Improved seeds will therefore be needed to significantly increase agricultural yields. The promotion of improved seeds to cover needs will provide more secure access and avoid waiting for government allocations, which are not made at the right time.

Innovative partnerships will be forged so that they can serve as the necessary guarantees for the granting of sufficient credit to relaunch profitable agricultural activity, particularly for the most vulnerable, i.e. young people and women. These are civil society organisations that are active in local development and rely on targeted aid to develop agricultural production resources.

Another factor to be taken into account is climate change, which is also being experienced in this commune, as in Burkina Faso and West Africa (IPCC, 2021). Agricultural activities are highly vulnerable to climate change, and diversification is a necessity (Bonkoungou, 2015). Diversification into other activities that are also vulnerable to climate change should be avoided. Climate-smart agriculture is a good alternative (Bayala et al., 2018; Hesse et al., 2013;

Hochet et al., 2012; Simpson, 2016).. Roads need to be built to open up the villages further west.

Farmers also use modern means of production. These include the use of herbicides and pesticides. Lacking the necessary financial and technical capacity, these modern production techniques are increasingly being adapted. But this adaptation is causing huge health risks due to the lack of protection (Congo, 2013; Gomgnimbou et al., 2010; Zan et al., 2023a, 2023b). Arboriculture, which is becoming increasingly popular, is one of these new production technologies. But the recommendations are scarcely followed and yields are below the expectations of the technical services. This also confirms hypothesis one.

There are a number of constraints undermining the rain-fed farming sector in this south-western part of Burkina Faso. The South-West is not a region favoured by the implementation of programmes. Where funding does exist, it is directed towards physical improvements that are often useful but beyond the reach of poor farmers. Cash crops, such as rice and cassava, attract more funding than food crops.

As a result, there is something of a disconnect between the farming practices implemented and the programmes and projects implemented. As a result, agricultural yields are low despite the fact that the region has very rich land.

The second hypothesis, which stipulates that farmers are encountering enormous difficulties during the start of the growing season, has also been verified. Solutions need to be found to enable this region with its very high potential to develop modern agriculture. Markets, of which there were only two when we visited, need to be developed. Practicable rural tracks will be able to facilitate these exchanges.

This will be a good opportunity to put climate-smart farming into practice, for agro-ecological farming that respects the environment and is sustainable for future generations. The local authorities will have the onerous task of changing their paradigm to put the interests of their constituents, the farmers, at the centre of their actions.

References

AGRECO. (2006). *"Environmental Profile of Burkina Faso.* http://ec.europa.eu/development/icenter/repository/burkina_faso_CEP_2006.pdf

Baiou, R. (2008). Africa and climate change. In *Changements climatiques: impacts, adaptation, mitigation* (p. 71).

Bayala, J., Dayamba, D. S., Ky-Dembele, C., Savadogo, P., & Arinloye, A. D. (2018). *Climate smart agriculture extension manual. Technical manual.* https://apps.worldagroforestry.org/downloads/Publications/PDFS/2018037.pdf

Bonkoungou, J. (2007). *Zoning des potentialités agricoles du bassin versant pilote du Zondoma, Burkina Faso* (Issue 226). Centre Régional AGRHYMET, Niamey, Niger.

Bonkoungou, J. (2015). *Climate variability, climate change and population vulnerability in Burkina Faso.* Abdou Moumouni University.

Burkina Faso. (2013). *Burkina Faso's National Sustainable Development Policy.* http://www.environnement.gov.bf/files/PNDD_Version_finale_du_17__10__2013.pdf

Camilla Toulmin, B. G. (2003). Transformations in West African agriculture and the role of family farms. *Development, 123*(1), 106.

CILSS. (2012). *Good agro-sylvo-pastoral practices for sustainable soil fertility improvement in Burkina Faso.*

Compaore, J., Bonkoungou, J., & Kiemdé, S. (2023). Effects of Communication for Better Vegetable Production in Burkina Faso: Case of the Agricultural Plain of Mogtedo in the Province of Ganzourgou in the Central Plateau Region. *EJFOOD, 5*(5), 5-11.

Congo, A. K. (2013). *Health risks associated with the use of pesticides around small reservoirs: the case of the Lombila dam* [2iE]. http://documentation.2ie-edu.org/cdi2ie/opac_css/doc_num.php?explnum_id=1825

United Nations Framework Convention on Climate Change. (2015). Paris Agreement. In *21st Conference of the Parties.* https://doi.org/10.1016/j.cub.2017.03.006

Diallo, B. (2010). *Perceptions endogènes, analyses agroclimatiques et stratégies d'adaptation aux variabilités et changements climatiques des populations dans trois zones climatiques du Burkina Faso* (Issue 226) [Centre Régional AGRHYMET, Niamey, Niger]. http://hdl.handle.net/10625/45775

Dugu, P., Autfray, P., Djamen, P., Girard, P., Olina, J., Ouedraogo, S., & Vall, E. (2012). L ' agroécologie pour l'agriculture familiale dans les pays du Sud: impasse ou voie d'avenir? Le cas des zones de savane cotonnière de l'Afrique de l'Ouest et du Centre. In *Cirad*.

IPCC. (2014). *Climate Change 2014. Mitigating climate change. Summary for policy makers.*

IPCC. (2021). *Climate Change 2021 The Physical Science Basis.* https://www.ipcc.ch/report/ar6/wg1/downloads/report/IPCC_AR6_WG1_SPM_French.pdf

Gomgnimbou, A. P. K. G., Nianogo, A. J., & Millogo-Rasolodimby, J. (2010). De la logique d'occupation spatiale à l'émergence des risques environnementaux dans la zone sud-Soudanienne du Burkina Faso: cas de l'interaction entre le coton et l'élevage. *Innovation et Développement Durable Dans l'agriculture et l'agroalimentaire*, 10.

Hauchart, V. (2007). *Sustainability of cropping practices in the northern Volta catchment* (Issue 2).

Hesse, B. C., Anderson, S., Cotula, L., Skinner, J., & Toulmin, C. (2013). *Building climate resilience in the Sahel* (Issue July). IIED. http://pubs.iied.org/pdfs/G03650.pdf

Hochet, P., Norbert, C., & Ehess, E. (2012). Describing policies to combat desertification in terms of land governance. A case study from western Burkina Faso. *Sécheresse, 23*, 196-201. https://doi.org/10.1684/sec2012.0346

IED. (2014). Family farming and the fight against poverty. *AGRIPADE, 30*(2), 36.

Kambiré, G. (2023). *Dynamics of hydro-climatic risks and endogenous adaptation strategies in the commune of Nako, Burkina Faso.*

MJE. (2007). *Etude sur les créneaux porteurs d'emplois, région du Nord, Burkina Faso.*

Naqvi, S. M. K., & Sejian, V. (2011). Global climate change: role of livestock. *Asian Journal of*

Agricultural Sciences, 3(1), 19-25.

Ouédraogo, I., Bonkoungou, J., & Yanogo, I. P. (2022). Climate-smart agriculture in a context of climate change and variability in Sub-Saharan Africa. *Djiboul, 3*(004), 500-515. https://medium.com/@arifwicaksanaa/pengertian-use-case-a7e576e1b6bf

Palé, S. (2020). *Morphohydrology of the Poni catchment (South-West Burkina Faso).* Joseph Ki-Zerbo.

Pouliot, M., Ouédraogo, B., Smith-Hall, C., & Simonsen, H. (2016). *Household-level studies of forests and poverty in Burkina Faso : Contextual information , methods and preliminary results* (Issue 47).

Simpson, B. M. (2016). *Preparing smallholder farm households to adapt to climate change. Pocket guide no. 1. Extension practice for agricultural adaptation.*

Slimane, A. Ben (2008). *Vulnerability of rainfed agriculture in the Volta Basin: analysis of the impact of rainfall variability and drought risk on yield and farm management* (Issue 13). IRD.

Somda, J., Issa, S., Moumini, S., Moussa, A. S., Goama, N., Josias, S., Silimana, B., Oumou, S. A., & Some, L. (2014). *Participatory vulnerability analysis and climate change adaptation planning in Yatenga, Burkina Faso* (No. 64). www.ccafs.cgiar.org

Sultan, B., Alhassane, A., Barbier, B., Baron, C., Tsogo, M. B., Berg, A., Dingkuhn, M., Fortilus, J., Kouressy, M., Leblois, A., Marteau, R., Muller, B., Oettli, P., Quirion, P., Roudier, P., & Traoré, S. B. (2012). The issue of vulnerability and adaptation of Sahelian agriculture to climate within the AMMA programme. In *La météorologie, Spécial AMMA* (pp. 64-72).

Tientiga, G. O., Zoundi, M., & Bonkungou, J. (2024). Impacts environnementaux et socio-économiques des Bassins de Collecte des Eaux de Ruissellement (BCER) Ecologiques , en irrigation d ' appoint : Cas de la zone d ' intervention du Projet BEOG PUUTO [Environmental and socio-economic impacts of ecologica. *International Journal of Innovation and Applied Studies, 43*(2), 513-541.

EU-AU. (2008). *Enhancing economic growth and reducing poverty: an African perspective.*

UNEP. (2008). *Africa, atlas of a changing environment.*

UNFCCC. (2021). *Nationally determined contributions under the Paris Agreement Synthesis report by the secretariat.* http://unfccc.int/resource/docs/2009/cop15/eng/11a01.pdf

United Nations - Reducing Emissions from Deforestation and forest Degradation (UN-REDD). (2015). *Developing a green economy in Africa: why forests matter.*

Yaméogo, A. (2015). *Impacts of gold panning on vegetation and soils in Perkoa, Sanguié province.* Ouagadougou.

Zan, A., Bonkoungou, J., Compaore, J., Sawadogo, B., & Kopola, R. (2023a). Impact of climate change on ruminant livestock production in the Lake Bam watershed in Burkina Faso. In M. Waziri Mato (Ed.), *Défis et perspectives de développement au Sahel : dynamiques environnementale, sociale et économique, conjonctures géopolitiques, crises sécuritaires et sanitaires* (pp. 201-210).

Zan, A., Bonkoungou, J., Compaore, J., Sawadogo, B., & Kopola, R. (2023b). Impact of climate change on ruminant livestock production in the Lake Bam watershed in Burkina Faso. *ACTES DU COLLOQUE HOMMAGES, TEMOIGNAGES ET RECONNAISSANCES AU Pr Tanga Pierre ZOUNGRANA, 3,* 201-210.

I want morebooks!

Buy your books fast and straightforward online - at one of world's fastest growing online book stores! Environmentally sound due to Print-on-Demand technologies.

Buy your books online at
www.morebooks.shop

Kaufen Sie Ihre Bücher schnell und unkompliziert online – auf einer der am schnellsten wachsenden Buchhandelsplattformen weltweit! Dank Print-On-Demand umwelt- und ressourcenschonend produziert.

Bücher schneller online kaufen
www.morebooks.shop

info@omniscriptum.com
www.omniscriptum.com

Printed by Books on Demand GmbH, Norderstedt / Germany